What's for lunch?

Peas

This edition 2003

Franklin Watts
96 Leonard Street
London
EC2A 4XD

Franklin Watts Australia
45-51 Huntley Street
Alexandria
NSW 2015

Editor: Samantha Armstrong
Series Designer: Kirstie Billingham
Designer: Jason Anscomb
Consultant: Processors and Growers Research Organisation
Reading Consultant: Prue Goodwin, Reading and Language
Information Centre, Reading

A CIP catalogue record for this book is available from the British Library
Dewey Decimal Classification Number 635

ISBN: 0 7496 4941 0

Printed in Hong Kong, China

What's for lunch?

Peas

Claire Llewellyn

W
FRANKLIN WATTS
LONDON • SYDNEY

Today we are having peas for lunch.

Peas are vegetables.

They contain **vitamins, minerals, fibre** and **protein.**

They help us to grow and stay healthy.

Peas are the seeds of the pea plant.
They grow inside cases called **pods.**
Most kinds of pea are taken out of their
pods, or shelled, before we eat them.
We don't shell **mangetout** and **sugarsnap** peas
but eat the pod with the peas inside.

sugarsnap
peas

6

Peas are grown all over the world. In spring, farmers plant the seeds in rows. This is called **drilling**. The seeds are peas from last year's **crop**.

In the soil the seeds split open and grow a **root** called a **radicle.** Then they sprout **shoots** that grow up through the soil. The young **seedlings** produce green leaves and side shoots, and grow into plants.

10

The farmers look after the growing crop. They spray it with **fungicide** and **insecticide** to protect it against diseases and insects.

In early summer, flowers grow on the pea plants. Soon the petals drop off and the pods begin to form. Inside each pod are up to ten tiny peas that start to swell. A month later, the peas inside the pods are sweet, juicy and green.

The peas are **harvested** by huge pea **viners**.
The machine takes the peas out of the pods.

Later the empty pods are dug back into the field as **fertilizer**.

The peas are poured into large containers and taken to nearby factories. Most of the peas we eat are **frozen.** The peas are frozen within three hours of being picked so they are really fresh.

In the freezing factory, the peas are graded or sorted into different sizes. They are cooked for just one minute, to kill any germs, and then blasted with ice-cold air. This freezes the peas and stops them from sticking together.

frozen peas

The frozen peas are packed into
bags and stored in freezers.
Refrigerated lorries transport them to shops.

Sometimes the peas are soaked in **brine** and
put into cans. This helps them to last longer.

Mangetout peas are grown in places where it is hot, like parts of Africa. When the peas are ready for harvesting, workers pick them and pack them into bags.

The bags are flown to many countries around the world. The peas are still crisp and fresh by the time they arrive.

mangetout
peas

2

Some peas are dried and split. They can be used to make soup.

Some peas are not harvested until they are hard. They have to be soaked in water before they are ready to eat.

split peas

Peas are eaten
in different ways around
the world. There are peas
in risotto, an Italian dish.

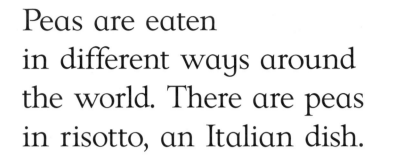

There are peas in pakora,
an Indian snack.

Peas are also
added to paella,
a Spanish dish.
Peas are sweet,
tasty and good
for you too.

29

Glossary

brine salty water which stops food decaying and helps it to last

crop what farmers grow in their fields

drill to plant seeds in rows

fertilizer something that helps plants grow

fibre something that helps us to digest our food

frozen when something is so cold it is hard and stays fresh

fungicide something that kills plant disease

to harvest to take the crop from the fields

insecticide something that kills insects

mangetout pea a kind of pea that is eaten in its pod

minerals materials found in rocks and also in our food. Minerals help us to stay healthy

pods the cases in which peas grow

protein	something found in food such as peas which helps to build and repair the body
radicle	the first root from the pea seed
root	the part of a plant that grows underground and takes moisture and goodness from the soil
seedlings	very young plants that grow from a seed
shoots	the new parts of a plant that grow above the soil
to shell	to take the peas out of their pods
sugarsnap pea	a kind of pea that is eaten in its pod
viner	a machine which picks pea plants and removes the peas from the pods
vitamins	something found in vegetables and fruit that keeps us healthy

Index

brine 22

crop 9

drill 9

fertilizer 17
fibre 5
frozen 18
fungicide 12

harvest 16, 25, 27

insecticide 12

mangetout peas 6, 25
minerals 5

pods 6, 14, 16, 17
protein 5

radicle 10
root 10

seedlings 10
shelled 6
shoots 10
sugarsnap peas 6

viner 16
vitamins 5

Picture credits: 7 Chris Fairclough, © Franklin Watts; 8, 9, 10 Holt Studios International / Nigel Cattlin; 11 Courtesy of Birds Eye Wall's Ltd; 12-13, 15 Holt Studios International / Nigel Cattlin; 16-17 Courtesy of FMC Harvesters; 19 Holt Studios International / Willem Harinck; 21, 23 Courtesy of Birds Eye Wall's Ltd; 24 Panos Pictures / Ron Giling. Cover Steve Shott; All other photographs Tim Ridley, Wells Street Studios, London.
With thanks to Aiden Senior and Alex Wright.